Table of Contents

1 Introduction to Ocean Petrology

Ocean petrology is essentially the study of rocks found beneath the waves. It's a fascinating field, combining the thrill of exploration with the meticulous analysis of Earth's processes. Unlike continental petrology, which focuses on land-based rocks, ocean petrology deals with a unique environment shaped by immense pressure, constant water interaction, and the dynamic movements of tectonic plates. Think about the sheer variety: from the volcanic basalt that forms the majority of the ocean floor to the unique minerals created near hydrothermal vents, and the layers of sediment built up over millennia. It's a world of contrasts, and understanding these contrasts is key to unraveling the mysteries of our planet's history.

Unlike examining rocks on land, where you can easily access outcrops, studying ocean floor rocks requires specialized technology and techniques. We often rely on remotely operated vehicles (ROVs) and autonomous underwater vehicles (AUVs) to collect samples. These submersibles are equipped with tools to collect rock samples, capture images, and even conduct in-situ analyses. Once these precious samples are brought back to the surface, detailed laboratory analyses are conducted, involving microscopic examination, chemical analyses, and isotopic dating. This allows us to determine the mineral composition, the formation age, and the environment in which the rocks originated.

The ocean floor isn't uniformly composed. Its petrological diversity reflects the complex geological processes at play. Mid-ocean ridges, for instance, are sites of intense volcanic activity where new oceanic crust is continuously formed. This crust, primarily composed of basalt, provides a rich source of information about the Earth's internal dynamics and mantle composition. In contrast, areas farther from the ridges often show signs of significant alteration due to interactions with seawater, leading to the formation of unique mineral assemblages and chemical signatures. These altered rocks offer a window into the long-term chemical exchange between the oceanic crust and the vast ocean.

Consider the fascinating case of hydrothermal vent systems. These are not just geographically distinct; they're petrologically unique. The intense heat and chemical reactions near these vents lead to the precipitation of unusual minerals, often rich in metallic sulfides. Analyzing these minerals allows us to understand the chemical processes that are occurring deep within the Earth's crust, providing insights into the conditions and chemical interactions that shape these unique ecosystems. The study of these vent minerals goes beyond simple composition. We can deduce temperatures, pressures, and fluid compositions that prevailed during their formation.

Sedimentary processes play a significant role in the ocean's petrology. The constant rain of particles from the surface waters, including biological remains and inorganic dust, accumulates on the seafloor. Over time, these particles undergo compaction and cementation, forming sedimentary rocks like limestone and chert. The composition of these sediments can tell us about past climate conditions, ocean currents, and biological productivity. Analyzing the layered structures within these sediments allows scientists to reconstruct ancient environments and the history of oceanographic changes over millions of years.

Understanding ocean petrology isn't merely an academic exercise. It has significant implications for various fields. Resource exploration, for instance, relies heavily on understanding the distribution of mineral deposits on the ocean floor. Studying the composition and formation of these deposits helps guide the search for valuable resources. Furthermore, knowing the petrological nature of the seafloor is crucial for assessing the environmental impact of activities like deep-sea mining. The potential disruption to ecosystems and the long-term consequences require a thorough understanding of the geological context, therefore highlighting the importance of ocean petrology. It's a vibrant field with practical applications and a multitude of intriguing discoveries yet to be made. The ocean floor remains a frontier; ocean petrology provides the tools to explore it.

1.1 Scope of Ocean Petrology

Ocean petrology isn't just about identifying rocks; it's about understanding Earth's dynamic processes. We examine the ocean floor's diverse rock formations—from the volcanic basalt of mid-ocean ridges to the layered sediments blanketing vast plains. This understanding helps us unravel the planet's geological history, from plate tectonics to past climates. Studying these rocks allows us to piece together the puzzle of how our planet has evolved, a journey spanning millions of years. It involves analyzing their mineral composition, textures, and the conditions under which they formed, giving us clues to the processes that shaped the ocean floor. Think of it like reading Earth's geological diary, written in stone.

The scope extends to understanding how these rocks interact with seawater, influenced by hydrothermal vents. These vents, often found near mid-ocean ridges, spew superheated, mineral-rich fluids, creating unique ecosystems. The chemical reactions between the rocks and these fluids alter the rocks' composition and contribute significantly to the overall ocean chemistry. This alteration process is a key area of research, helping us understand the cycling of elements within the Earth system. It also helps us understand how these hydrothermal systems might potentially harbor life in extreme conditions.

Moreover, ocean petrology plays a vital role in mapping the ocean floor. By analyzing rock samples obtained through various techniques, including drilling and submersible exploration, we can build a comprehensive understanding of the underlying structures. This knowledge has implications for understanding seafloor spreading, subduction zones, and the distribution of valuable resources, from polymetallic nodules to geothermal energy potential. The rocks tell stories of past volcanic eruptions, tectonic shifts, and even ancient life forms.

Furthermore, understanding the interplay between magmatic processes and the formation of oceanic crust is a central theme. The creation of new crust at mid-ocean ridges, a key aspect of plate tectonics, is fundamentally linked to the petrology of the basalts that form there. This process involves the melting of the Earth's mantle, the ascent of magma, and its subsequent cooling and crystallization. Analyzing these basalts provides crucial information about the composition of the Earth's mantle and the dynamics of plate boundary processes.

Consider the sedimentary rocks. These aren't just passive layers; they record a history of ocean currents, climate change, and biological activity. Their composition—from the tiny skeletons of marine organisms to windblown dust—reveals much about past ocean conditions. By carefully studying their layered structures, scientists can reconstruct past environmental changes. This helps us understand long-term climate trends and how the oceans have responded to these changes. Think about how the composition of sediments reflects past marine life and can help in understanding the evolution of species and even past ocean temperatures.

The significance also expands into the realm of resource exploration. The ocean floor is rich in various resources, including manganese nodules, cobalt crusts, and potentially even rare earth elements. Understanding the petrology of these deposits is critical for efficient and environmentally responsible extraction. This requires careful planning and the development of sustainable practices. It's not just about finding resources; it's about managing them responsibly for future generations.

Finally, remember that ocean petrology is a continuously evolving field. New technologies, such as advanced imaging techniques and sophisticated analytical methods, continually refine our understanding of the ocean's geological processes. This field integrates knowledge from multiple disciplines, such as geochemistry, geophysics, and biology. This interdisciplinary approach provides a more holistic picture of the complex interactions that shape the ocean floor. Ongoing research continually enhances our capacity to understand Earth's dynamic systems and protect its valuable resources. The ocean floor continues to hold numerous mysteries, making the future of ocean petrology both exciting and vital.

1.2 Key Petrologic Concepts

Imagine the ocean floor, a vast, unexplored landscape. It's not just a flat expanse; it's a dynamic realm shaped by powerful geological forces, resulting in a stunning array of rock types and structures. Understanding these requires grasping some key petrologic concepts. We need to understand how these rocks form, what they're made of, and how they change over time. This chapter provides you with that essential foundation.

Firstly, let's define what we mean by "igneous rocks." These rocks are formed from the cooling and solidification of molten rock, or magma. In the ocean, most igneous rocks are basalts, formed by volcanic activity at mid-ocean ridges. These ridges are the underwater mountain ranges where tectonic plates pull apart, allowing magma to rise and erupt. The rate of cooling significantly impacts the texture of the resulting rock; rapid cooling produces fine-grained basalt, while slow cooling leads to coarser-grained varieties. Consider how different the textures might be between a lava flow rapidly cooling on the ocean surface and a magma chamber slowly crystallizing deep beneath the seafloor. These differences reflect different geological processes at work.

Next, think about mineral identification. Rocks are composed of minerals, and identifying these components is fundamental to petrology. We use various techniques, including microscopic analysis and chemical testing, to determine the mineral composition of a rock sample. Common minerals in oceanic rocks include plagioclase feldspar, pyroxene, and olivine. The specific proportions of these minerals, along with their textures and other features, help us classify the rock and infer its formation history. It is like a detective story; each mineral clue hints at the conditions under which the rock formed—temperature, pressure, and the availability of different elements.

Understanding mineral structures is paramount. Minerals have characteristic crystal structures that dictate their physical and chemical properties. These structures influence a rock's overall properties such as strength, density, and reactivity with seawater. For example, the crystal structure of olivine makes it relatively susceptible to alteration by hydrothermal fluids, a process we'll discuss later. Learning to recognize these structures is like learning a new language, opening doors to understanding the complex interactions between minerals and their environment.

Now, let's talk about sedimentary rocks. Unlike igneous rocks, which form from molten material, sedimentary rocks are formed from the accumulation and cementation of sediments. In the ocean, sediments can be of various origins: biogenic (from marine organisms), terrigenous (from land sources), or volcanogenic (from volcanic eruptions). The type of sediment and the environment of deposition both affect the resulting rock's characteristics. Deep-sea sediments are often fine-grained clays and silts, while shallower areas may contain coarser-grained sands and gravels. These differ-

ences reveal much about past ocean currents and sea levels.
Furthermore, the concept of metamorphic rocks is crucial. These are rocks that have been transformed by heat, pressure, or chemical reactions, often during tectonic events. While less common in the pristine oceanic crust, metamorphic rocks can be found near subduction zones or in areas of intense hydrothermal activity. The metamorphism of basalt, for instance, can produce rocks like greenschist or amphibolite, reflecting changing pressure and temperature conditions within the Earth's crust. Recognizing these metamorphic changes allows us to reconstruct the geological history of a particular area. These changes leave clear marks, like fingerprints in stone, indicating the intense events that shaped the rock.
Finally, consider the concept of rock alteration. This encompasses the changes a rock undergoes after its initial formation. Hydrothermal alteration, driven by hot, chemically active fluids circulating through the crust, is particularly significant in the ocean environment. These fluids can dissolve or replace minerals, changing the rock's composition and properties. Weathering, the breakdown of rocks at the Earth's surface, is another important alteration process. Understanding alteration is vital because it can significantly impact the rock's properties and interpretation of its initial formation. It's like comparing a pristine painting to one exposed to the elements over time; the original beauty remains, but the changes are evident. This is true for oceanic rocks.
By understanding these key petrologic concepts—igneous, sedi-

mentary, and metamorphic rock formation, mineral identification, mineral structures, and rock alteration—you'll be well-equipped to explore the diverse and fascinating rock formations of the ocean floor. This knowledge provides a strong foundation for interpreting the data obtained from mapping technologies and for understanding the complex geological processes that shape our planet's oceans.

2 Mapping the Ocean Floor

Imagine the ocean floor, a vast, unexplored landscape hidden beneath the waves. To understand it, we need maps, detailed charts that reveal its hidden secrets, its mountains and valleys, its plains and canyons. This isn't as simple as drawing a line on a piece of paper; mapping the ocean floor requires sophisticated technology and ingenious methods.

Sonar, short for sound navigation and ranging, is one of our most important tools. It works by sending sound pulses into the water; these pulses bounce off the ocean floor and other underwater objects, returning to a receiver. By measuring the time it takes for these echoes to return, scientists can calculate the distance to the seabed, creating detailed bathymetric maps. The precision of sonar varies depending on the frequency of sound used, with higher frequencies providing more detail but penetrating less deeply into the water column. Think of it like shining a flashlight—a narrow beam gives a clearer, more focused view, but doesn't illuminate as much area.

Beyond simple depth measurements, seismic reflection surveys offer a glimpse beneath the surface. These powerful techniques use sound waves of much lower frequencies than sonar to penetrate far deeper into the Earth's crust. These sound waves reflect off boundaries between different layers of rock and sediment, revealing the subsurface structure of the ocean floor. Imagine seeing not only the surface of the land, but also the underground layers, creating a three-dimensional picture of its composition. This is essential for understanding the geological history of the area, identifying potential resources and predicting geological hazards.

Remote sensing, although not directly mapping the ocean floor itself, still provides valuable information. Satellite altimetry, for instance, measures subtle variations in sea surface height caused by the gravitational pull of underwater mountains and trenches. This technique allows scientists to detect large-scale features on the seafloor, providing a broad context for higher-resolution surveys. This is a great method for covering vast areas quickly, identifying areas for further investigation using more detailed techniques. Think of it as getting a bird's-eye view before zooming in for a closer look.

Different sonar systems use various frequencies. Multibeam sonar, for example, sends out a fan of sound waves, creating a swath of data with each pass. This significantly speeds up the mapping process compared to older single-beam systems. Side-scan sonar, on the other hand, "sees" to either side of the vessel's path, detecting objects on the seafloor and in the water column. The choice of sonar system depends on the specific goal of the mapping project, the depth of water, and the level of detail required. Essentially, selecting the right tool is as crucial as the skill of its operator.

Further enhancing our understanding are advanced techniques. Autonomous underwater vehicles (AUVs) and remotely operated vehicles (ROVs) are robotic explorers capable of detailed surveys in challenging environments. AUVs, free-ranging underwater robots, can cover large areas autonomously, while ROVs, tethered to a surface vessel, allow for real-time control and the collection of samples. These robots, equipped with a variety of sensors, cameras, and manipulators, provide a unique combination of reach and precision, allowing for close-up examination of interesting geological features. This is akin to exploring a previously uncharted territory with the flexibility to adjust and delve deeper into specific areas.

Three-dimensional models, synthesized from data gathered through these various methods, provide a remarkably realistic representation of the ocean floor. These models are not merely visual aids; they are powerful tools used for geological interpretation, resource exploration, and environmental monitoring. They allow scientists to visualize complex geological structures, analyze sediment distribution, and understand the interactions between various geological processes. Imagine having a virtual underwater world at your fingertips, providing all the needed information for study and re-

search.
The ongoing development of new technologies promises to accelerate the pace of ocean floor mapping. More advanced sonar systems, improved sensor technologies, and more sophisticated data processing techniques will reveal ever greater detail about this hidden world. The future of ocean floor mapping lies in the integration of multiple data sources, allowing for a more holistic understanding of the ocean floor. The future is bright, a constant discovery of new wonders, and a journey of infinite knowledge to be gained.

2.1 Sonar Technology

Sonar, short for sound navigation and ranging, is a pivotal tool in mapping the ocean floor. Imagine sending sound waves into the water; they bounce off the seafloor and other underwater objects, returning to a receiver. The time it takes for the sound to travel and return directly relates to the distance – it's a clever application of basic physics. This technique allows us to create detailed images of the ocean's depths, revealing previously unseen landscapes and geological structures.
The development of sonar technology wasn't a single breakthrough, but rather an evolution. Early sonar systems were rudimentary, offering only basic depth readings. As technology advanced, so did the resolution and sophistication of sonar systems. Modern sonar uses a variety of frequencies, each capable of penetrating the water at different depths and providing different levels of detail. High-

frequency sonar excels at providing fine-scale imagery of the immediate seafloor. Low-frequency sonar, on the other hand, penetrates much deeper, providing insights into the structure of the seafloor beneath the sediment layers.

Think of it like this: high-frequency sonar is like a magnifying glass, showing intricate details of a small area. Low-frequency sonar, however, is more like a broad-spectrum X-ray, revealing the larger, hidden structures beneath the surface. This allows scientists to build a comprehensive picture of the ocean floor, from the fine grain sizes of the sediment to the massive shapes of underwater mountains and canyons.

One crucial aspect is side-scan sonar. This type of sonar doesn't just measure depth; it also "scans" the ocean floor to the sides of the sonar source. This provides a visual representation of the seafloor, similar to an aerial photograph, showcasing variations in texture, features, and even the presence of objects on the seabed. This is particularly useful in searching for shipwrecks, geological anomalies, or other features of interest that might be missed using only traditional depth-measuring sonar. It's basically a sonar "camera."

Multibeam sonar, a significant advancement, employs numerous sound beams simultaneously, painting a far broader swathe of the seabed with each pass. Instead of a narrow line of data, it creates a wide, detailed image, drastically speeding up the mapping process and enabling the creation of high-resolution maps. This technique has dramatically improved our ability to map the extensive ocean floor, providing an unprecedented level of detail compared to older single-beam sonar systems. Think of it like upgrading from a pencil sketch to a vibrant, high-definition photograph.

Furthermore, the data collected from sonar is not just used for simple mapping. Sophisticated software processes the sonar data, creating three-dimensional models of the seafloor. These models are crucial for understanding geological processes, such as the formation of submarine volcanoes, the movement of tectonic plates, and the distribution of various sediments. The data can also aid in understanding the dynamics of underwater currents, as the shape of the seafloor directly influences water flow.

Beyond these applications, sonar is invaluable in the search for underwater resources. Oil and gas exploration relies heavily on sonar to locate potential hydrocarbon reserves hidden beneath the seafloor. Similarly, the search for valuable minerals and other resources on the ocean floor makes use of the information provided by sonar technology. Accurate mapping of the seafloor is thus crucial for efficient and responsible exploration and extraction. This makes sonar not only a tool for scientific discovery, but also a key element in resource management and economic activity in the marine environment.

Even the study of marine life benefits from sonar technology. By analyzing sonar reflections, scientists can study the distribution and behavior of marine animals, gaining insights into their habitats, migration patterns, and interactions within the marine ecosystem. This understanding is crucial for conservation efforts and for managing human impact on the oceans. So, sonar's contribution goes far beyond merely creating maps; it's a crucial tool in understanding the whole ocean ecosystem.

In conclusion, sonar technology has profoundly transformed our understanding of the ocean floor. From basic depth measurements to the creation of detailed 3D models, sonar has allowed us to explore and map a previously unseen world. Its applications range from fundamental scientific research to resource exploration and environmental management, showcasing its importance in various fields connected to oceanographic studies. The continued development and refinement of sonar techniques will undoubtedly lead to even more exciting discoveries and advances in our knowledge of this dynamic and fascinating underwater realm. It's a testament to human ingenuity and our enduring quest to unravel the mysteries of our planet.

2.2 Seismic Reflection

Seismic reflection is like looking into the Earth's past, using sound waves instead of a time machine. Imagine sending sound pulses deep into the ocean floor; some waves bounce back from different layers of sediment and rock, creating echoes. These echoes, recorded by sensitive receivers, are then used to build a picture of the subsurface structure, revealing hidden layers of different densities, akin to a detailed geological X-ray. This technique helps us understand the layers of the seafloor, offering clues to past geological events and the composition of the ocean crust. We're not just looking at the surface; we're peeling back the layers, uncovering secrets buried beneath.

The process starts with a powerful sound source, often an air gun or a sparker, generating sound waves that travel through the water column and penetrate the seafloor. These waves travel at varying speeds depending on the density and elastic properties of the materials they encounter, encountering changes in these properties, like boundaries between layers of sediment or rock, and at each interface some sound energy is reflected back. The reflected waves travel back to the surface where specialized hydrophones or geophones, incredibly sensitive microphones designed to detect underwater sounds, act as the "ears" of our system, meticulously recording the returning echoes.

Analyzing these echoes isn't a simple task; it demands sophisticated processing techniques. The time it takes for the sound waves to travel down and back up is directly related to the depth of the reflecting layer. By meticulously measuring these travel times, scientists can construct a seismic profile, a cross-section of the subsurface layers, essentially a detailed "map" of what's underneath the ocean's floor. This involves specialized software and algorithms that account for various factors like water depth, wave propagation, and noise interference. The result is a remarkably detailed image of the ocean floor's hidden architecture.

Think of it like a layered cake: each layer has a different texture and density. Seismic reflection allows us to discern the individual layers— their thickness, composition, and even the subtle differences within them. We can identify various sediment types, the boundaries between sedimentary rocks and igneous rocks like basalt, and perhaps even locate faults and other geological structures which greatly influence the features of the ocean floor. The resulting images give us valuable insights into the history of the ocean floor, allowing us to trace the movements of tectonic plates and understand the processes that shaped the ocean basins over millions of years.

Beyond simple layer identification, seismic reflection can reveal much more. The strength and clarity of the reflections reveal information about the nature of the interfaces between the layers. A sharp reflection often indicates a distinct boundary between two layers of significantly different properties. In contrast, a diffuse or weak reflection might suggest a more gradual transition between layers, or the presence of porosity or fractures within a layer. This detailed information is crucial in understanding the geological processes and helps geoscientists distinguish between sedimentary layers with different compositions, for example.

One impressive application of seismic reflection is in the exploration of subsurface resources. Identifying potential hydrocarbon reservoirs beneath the ocean floor is just one example. By analyzing the seismic data, geologists can look for specific geological formations that are known to trap oil and gas. Similarly, it can help locate mineral deposits and assess their size and extent, providing invaluable information for resource management and exploration. It is not only about extracting resources, though; understanding the subsurface structure is critical for responsible resource management, allowing a proper assessment of environmental impacts and risks.

Furthermore, the technique is essential for understanding the processes that shape the ocean floor. It allows researchers to visualize the extent of submarine landslides and understand their impact on the environment. Mapping these hazards helps in mitigating risks and developing strategies for coastal protection and safety. Moreover, seismic reflection assists in studying the formation and evolution of hydrothermal vent systems, those remarkable deep-sea ecosystems thriving around hot springs. By studying the subsurface structure around these vents, we can learn more about their fluid circulation, chemical processes, and geological evolution, which gives insight into their formation, stability, and evolution over time.

Seismic reflection is a powerful tool, a window into the secrets hidden beneath the ocean's surface. It allows us to explore the Earth's history, understand geological processes, and discover valuable resources. But it's not just about discovery; it's also about responsible stewardship, informing sustainable practices and protecting our oceans for future generations. By continuously refining techniques and integrating them with other data sources such as those gathered with autonomous underwater vehicles (AUVs) and remotely operated vehicles (ROVs), our knowledge of the ocean depths continues to expand and our understanding of Earth's intricate systems becomes evermore precise. The future of marine geological exploration holds exciting possibilities, unlocking the mysteries of our planet's hidden depths and shaping our understanding of its profound processes. This detailed imaging of the seafloor contributes significantly to a greater understanding of the dynamic processes that shape our planet.

2.3 Remote Sensing Techniques

Remote sensing, from its vantage point above the waves, offers a unique perspective on the ocean floor. Unlike sonar which relies on direct sound transmission, remote sensing uses various electromagnetic wavelengths – from visible light to microwaves – to gather data. Imagine it as taking a picture of the ocean, but instead of just seeing the surface, we're peering beneath, albeit indirectly. This method provides a broader-scale view, ideal for mapping large areas quickly, supplementing the detailed information provided by techniques like sonar. Data collected, processed and interpreted from these remote observations is crucial to understanding the large-scale patterns of the ocean floor.

Satellite altimetry, one of the key remote sensing tools, measures the subtle variations in sea surface height. These variations, often just centimeters, are surprisingly informative. They aren't caused by waves directly, but by the gravitational pull of underwater features. A massive underwater mountain, for instance, will exert a stronger gravitational pull, causing a slight bulge in the sea surface above it. By meticulously measuring these subtle bulges, scientists can infer the shape and size of features far below. It's like feeling the weight of a hidden object through its subtle influence on its surroundings. This technique is particularly useful for mapping large, broad features such as the vast mid-ocean ridges.

Another invaluable tool is radar altimetry. Radar systems transmit microwave pulses towards the sea surface. The time it takes for the pulses to return, as well as the strength of the reflected signal, provides details about the sea surface roughness and height. In turn, this reveals information about the underlying seafloor. Areas with rougher seas might indicate shallower depths or strong currents, influenced by seafloor features. Similarly, variations in the microwave backscatter, how much of the signal is returned, can indicate differences in seafloor composition. For example, areas with significant volcanic activity might exhibit different backscatter patterns due to alterations in the water column above hydrothermal vents.

Beyond altimetry, other remote sensing techniques contribute to ocean floor mapping. Satellite-based gravity measurements, for example, offer incredibly high resolution detail. These measurements show very minute changes in the Earth's gravitational field, providing insights into subtle variations in density within the Earth's crust and mantle beneath the ocean. This provides valuable insights into the structure of the ocean floor and its relationship with the deeper structures, even revealing the presence of previously unknown subsurface features. It's as if the ocean floor is subtly revealing its secrets through its gravitational pull.

Moreover, multispectral and hyperspectral imagery, usually employed from airborne platforms, offer spectral information about the water column itself. By analyzing different wavelengths of light reflected or absorbed by the water, we gain insight into water depth, turbidity, and the presence of certain sediments or minerals. For example, the presence of specific phytoplankton can indirectly indicate nutrient levels near the seafloor, aiding in understanding the distribution of certain sedimentary deposits. This spectral information is used to infer the composition and characteristics of the seafloor indirectly. It's like looking at the water's color to determine the type of soil beneath it – indirect but powerfully informative.

While remote sensing doesn't provide the fine-grained detail of direct methods like sonar or submersible surveys, it offers unparalleled coverage across vast distances. It provides a crucial context for other mapping techniques, guiding where to focus more intense investigation. Remote sensing is not a stand-alone solution, but a powerful complementary tool that significantly enhances our ability to understand and map the vast and enigmatic ocean floor. Combining this technique with others provides a powerful combination of broader-scale mapping capabilities and high-resolution detailed insights. It's truly a remarkable tool that allows us to effectively explore the hidden world below the waves. The future of ocean mapping surely incorporates remote sensing for both its speed and scope. The more we utilize and advance these technologies, the more of the ocean's secrets will be revealed.

3 Hydrothermal Vent Systems

Imagine the ocean floor, not as a flat, sandy expanse, but a landscape teeming with life, fueled not by sunlight, but by the Earth's own internal heat. These are hydrothermal vent systems, oases of life in the otherwise barren deep. They form where tectonic plates meet, allowing seawater to seep down into the crust, heated by magma. This superheated water, rich in dissolved minerals, then erupts back out, creating spectacular chimney-like structures and supporting bizarre ecosystems. The vents' formation is a fascinating interplay of geology and chemistry; imagine the pressure, the heat, the incredible chemical reactions creating these underwater wonders.

Water circulating through the oceanic crust undergoes dramatic changes. Its temperature increases dramatically, reaching hundreds of degrees Celsius. As it does so, it dissolves various minerals from the surrounding rocks, becoming a chemical cocktail vastly different from ordinary seawater. This hydrothermal fluid is rich in dissolved metals, like iron, copper, and zinc, and gases, like hydrogen sulfide. It's a unique chemical environment, not just in temperature but in the presence of these elements which are toxic to most life forms – yet, here life thrives.

The chemistry of hydrothermal vents is incredibly complex, involving a range of chemical reactions and mineral precipitation. The dissolved minerals in the vent fluid precipitate out as the water cools and mixes with the surrounding seawater, creating those distinctive mineral deposits we see forming the vents themselves. Scientists are still trying to fully understand the precise chemistry, the subtle variations that occur between different vent systems, and the long-term evolution of these systems. It's a dynamic process, constantly changing and evolving, a testament to the planet's dynamic inner workings.

The life found around hydrothermal vents is nothing short of extraordinary. These ecosystems are chemosynthetic, meaning organisms derive their energy from chemical reactions rather than photosynthesis. Bacteria thrive on the chemicals in the vent fluid, forming the base of the food web. Giant tube worms, blind shrimp, and other unique creatures have adapted to this extreme environment, making their living from the energy-rich vent fluids. This is a testament to life's remarkable resilience and adaptability, a world entirely independent of sunlight.

Vent fluid circulation drives this entire system. The heated water rises from the vents, creating plumes that can extend hundreds of meters into the water column. These plumes carry the dissolved chemicals and provide a pathway for the transport of organisms and materials. The circulation patterns are complex and influenced by factors like the vent's location, the shape of the seafloor, and the surrounding ocean currents. Understanding this circulation is crucial for modeling the vent ecosystem and predicting its response to changes in the environment, whether natural or man-made.

The dynamics of a hydrothermal vent ecosystem are a delicate balance between the supply of chemicals from the vents and the consumption by the organisms that live there. Changes in the vent fluid flow, its chemical composition, or the populations of organisms can all have cascading effects on the entire ecosystem. Research into these ecosystems helps us understand how life can adapt to extreme environments and provides insights into the origins of life itself. The fragile balance reminds us of the importance of conservation and sustainable practices, especially given the increasing pressures of human activities on the ocean. Studying these amazing places offers valuable lessons about resilience, adaptability, and the interconnectedness of our planet.

3.1 Vent Formation

Deep beneath the ocean's surface, where sunlight fails to penetrate, lies a world of wonder and mystery: hydrothermal vents. These remarkable ecosystems thrive in the absence of photosynthesis, fueled instead by chemical reactions. Their formation, a complex interplay of geological forces, is a captivating story of Earth's dynamic processes.

Imagine the immense pressure deep within the Earth's crust. Molten rock, or magma, incredibly hot and under tremendous pressure, seeks release. Near mid-ocean ridges, where tectonic plates pull apart, this magma finds a pathway to the surface. The escaping magma isn't always a violent eruption; sometimes, it seeps slowly through cracks and fissures in the ocean floor, creating a network of underground pathways.

Seawater, drawn into these fissures by the heat and pressure gradients, percolates through the hot rocks. This water becomes superheated, dissolving minerals from the surrounding basalt. The resulting fluids, enriched with dissolved minerals and gases, are incredibly hot – sometimes exceeding 300°C (572°F) – despite the frigid surrounding water. This heated, mineral-rich water is significantly less dense than the surrounding ocean water.

Buoyancy plays a key role. As the superheated water, now much less dense, rises through cracks and fissures, it creates immense pressure building within the oceanic crust. This pressure eventually overcomes the surrounding rock's resistance. This is the genesis of the hydrothermal vent. When the pressure becomes too great, the superheated water bursts forth from the ocean floor. The dramatic expulsion of this mineral-rich fluid forms the iconic black smoker vents, billowing dark plumes into the surrounding cold ocean water. The minerals, rapidly cooling and precipitating out of solution, build up around the vent openings, creating the characteristic chimney-like structures.

The precise location of vents is highly dependent on geological factors. For instance, the proximity to active volcanic areas, such as mid-ocean ridges, is critical. These are areas of intense tectonic activity, characterized by the formation of new oceanic crust. The fractures and fissures created during this process provide pathways for magma and seawater interaction, generating the conditions necessary for vent formation. The specific type of vent – black smoker, white smoker, or diffuse flow – is also influenced by the chemistry of the hydrothermal fluids and their interaction with the surrounding seawater.

Understanding vent formation involves appreciating the crucial interplay between plate tectonics, magma genesis, and fluid dynamics. The geometry of the oceanic crust, including its fracture patterns and permeability, heavily influences the location, size, and activity of vents. Variations in the composition of the underlying basalt, the amount of seawater interacting with the magma, and even the flow rates of the hydrothermal fluids all affect the vent's characteristics. The chemical composition of the vent fluids is another critical factor, determining the types of minerals deposited and the unique chemical environment supporting life around them. Scientists are constantly working to refine our understanding of these complex interactions, improving our ability to predict and model the formation of these fascinating underwater ecosystems.

Over time, the continued activity of hydrothermal vents leads to the formation of significant mineral deposits on the ocean floor. These deposits frequently contain valuable metals, such as copper, zinc, and gold, making hydrothermal vents of great interest for future resource exploration. However, exploration must proceed cautiously to minimize the environmental impact and ensure the preservation of these unique ecosystems. The interplay of geology, chemistry, and biology, evident in the intricate formation and ecosystem of hydrothermal vents, underscores the remarkable complexity and interconnectedness of Earth's systems.

3.2 Chemistry of Vents

The water spewing from hydrothermal vents isn't just hot; it's a chemical cocktail. Imagine a geyser, but instead of plain water, it's a superheated solution brimming with dissolved minerals. These minerals, leached from the Earth's crust as the water percolates through the rocks, are the heart of the vent's chemistry. They dictate the vent's unique ecosystem and play a vital role in global geochemical cycles. Sulfides, like those of iron, copper, and zinc, are particularly abundant, often precipitating near the vent openings, creating those iconic, chimney-like structures we associate with hydrothermal vents.

Concentrations of these dissolved metals fluctuate wildly. Changes in pressure, temperature, and the rate of seawater circulation all influence the chemical composition. This dynamic environment makes predicting precise concentrations challenging, but it also makes studying these vents incredibly fascinating. Think of it like a natural laboratory, constantly experimenting and producing new combinations of chemical compounds. Each vent, even those close together, has subtly different chemical signatures, reflecting the complex interplay of geological and hydrological processes below the seafloor. The fluids themselves are not uniform throughout the vent system. They're constantly mixing and reacting, creating gradients and micro-environments that support a staggeringly diverse range of life.

One key element shaping vent chemistry is pH. The fluids emerging from vents are often highly acidic due to the presence of dissolved gases like hydrogen sulfide and carbon dioxide. As these fluids mix with the relatively alkaline seawater, the pH changes dramatically, influencing mineral precipitation and the survival of the organisms that thrive in this extreme environment. Furthermore, the chemical reactions between the vent fluids and the surrounding seawater aren't simply random; they follow well-established chemical principles. Oxidation-reduction reactions, for example, play a critical role in the cycling of elements like sulfur and iron, driving many of the chemical changes observed near vents.

The vent fluids themselves are rich in dissolved gases. Hydrogen sulfide (HS), a toxic gas to most life forms, is a key component, providing the energy source for chemosynthetic organisms – the foundation of the vent ecosystem. Methane and other hydrocarbons are also present, adding further complexity to the chemical mix. Measuring these gas concentrations, both in the vent fluids and the surrounding seawater, provides valuable insights into the geological processes occurring deep beneath the ocean floor. It's like solving a complex chemical puzzle; understanding the input, the reactions, and the resulting output helps us piece together the entire picture of vent activity.

Beyond the major components, a vast array of trace elements is also found in vent fluids. These trace elements, present in smaller quantities, can significantly impact the chemical and biological processes occurring at vents. For instance, the presence of certain rare earth elements might reflect the specific geological conditions at the vent site, providing clues to the formation and evolution of the underlying hydrothermal systems. Careful analysis of these trace elements requires advanced analytical techniques, but the insights they provide are invaluable for understanding the subtle variations between vents and the overall geological context. Scientists use sophisticated tools like mass spectrometry to precisely identify and quantify these trace components. The data reveal a surprisingly complex chemistry, far beyond the few major elements originally suspected.

Consider the impact on the surrounding ocean. Vent fluids are constantly interacting with the surrounding water, influencing the chemical composition of the deep ocean. The release of dissolved minerals and gases affects the overall oceanic chemistry, even on a global scale. Some researchers believe vents play a significant role in regulating the global carbon cycle, impacting climate and oceanic systems. Therefore, understanding the chemical nature of vents isn't just crucial for understanding the specific ecosystem, it's critical for grasping the larger picture of the earth's systems. The research also shows a surprising variability in the chemistry of different vents across various oceanic regions, making each vent a unique entity.

The study of vent chemistry isn't just about identifying individual components; it involves understanding the complex interactions between those components. It requires analyzing the dynamic processes that shape the chemical environment, considering factors such as temperature gradients, pressure changes, and fluid mixing. It's about building models that can simulate these complex interactions, predicting the behavior of these fascinating undersea ecosystems. These simulations are refined constantly, incorporating new data and improving our understanding of the dynamic nature of the vents.

Ultimately, unraveling the chemistry of vents is like piecing together a giant, intricate puzzle. Each piece – a mineral, a gas, a reaction – contributes to a broader understanding of these remarkable ecosystems and their role in the planet's complex interconnected systems. The more we learn about this chemistry, the deeper our appreciation for the intricate beauty and hidden power of the deep ocean. It's an ongoing process of exploration and discovery, constantly revealing new insights into the chemical wonders of the deep. And it's this constant learning and adapting that drives the science forward. Each research expedition contributes another piece, slowly revealing a more complete understanding of this hidden world.

3.3 Biology of Vents

Imagine a world without sunlight, a world where crushing pressure reigns and scalding water spews from the Earth's depths. This is the realm of hydrothermal vents, and within this seemingly inhospitable environment thrives a bizarre and fascinating ecosystem, entirely unlike anything we see on land. These vents, fueled by geothermal energy, aren't just geological features; they're vibrant biological hotspots, supporting a unique community of organisms. Their existence fundamentally alters our understanding of life's possibilities and the adaptability of life itself.

The foundation of this unique ecosystem lies in chemosynthesis. Unlike plants, which rely on photosynthesis to convert sunlight into energy, vent organisms depend on chemosynthetic bacteria. These remarkable microbes harness the energy released from chemical reactions, primarily the oxidation of hydrogen sulfide found in vent fluids. This process fuels a complex food web, supporting a diversity of life ranging from tubeworms and mussels to shrimp and fish, all uniquely adapted to survive in these extreme conditions.

Giant tubeworms, for instance, lack a digestive system entirely.

They rely entirely on symbiotic bacteria residing within their bodies to provide them with nutrients. These bacteria perform chemosynthesis, converting the chemicals from vent fluids into energy that nourishes their host. The worms' bright red plumes, waving in the vent's currents, are specialized structures that absorb the necessary chemicals. This symbiotic relationship highlights the remarkable interdependence within the vent community. Mussels, too, participate in this complex interplay, housing chemosynthetic bacteria within their gills, creating a similar energy-producing partnership.

The organisms found at hydrothermal vents demonstrate remarkable adaptations to their extreme environment. Many possess specialized enzymes capable of withstanding extreme temperatures and pressures. Others display unique metabolic processes, utilizing various chemicals in the vent fluids as their energy source. Their pigmentation, often deep reds and blacks, reflects their adaptation to the darkness of the deep sea. For example, some vent species have heightened senses, including chemoreception, enabling them to detect and locate vent fluids amidst the vast ocean expanse.

The distribution and abundance of vent communities aren't uniform. They are closely tied to the activity of the vent itself; the strength and chemistry of the vent fluids dictate the types and numbers of organisms found at a specific location. Vents with high-temperature fluids often support fewer species, while those with lower temperatures show higher biodiversity, showcasing the intricate relationships between geological processes and biological diversity. Studying these subtle differences helps unravel the complex dynamics of these vent ecosystems.

Understanding the biology of vents necessitates considering the dynamic nature of their environment. Vent fields are ephemeral, frequently altered by volcanic and tectonic activity. Eruptions and shifts in fluid flow can drastically impact vent communities, leading to both local extinctions and the creation of new habitats. This inherent instability necessitates remarkable adaptability among vent organisms, constantly adjusting to changes in temperature, chemical composition, and even physical location.

Vent ecosystems also face threats from human activities. Deep-sea mining, the extraction of valuable minerals from the seafloor, poses a significant risk to these fragile habitats. Pollution from industrial activities can also indirectly impact these remote ecosystems. Preservation efforts are crucial to protect these unique and vital environments. Their study provides invaluable insights into the fundamental principles of life on Earth, highlighting the remarkable resilience and adaptability of life in extreme conditions. The lessons learned from these deep-sea communities can inform our understanding of potential life in other extreme environments, perhaps even beyond our planet. Further research continues to unveil the secrets of this remarkable world, enriching our understanding of life's tenacity and the planet's remarkable biodiversity. The deep ocean floor remains a treasure trove of discoveries, and the biology of vents is a testament to the planet's wondrous and unexpected potential for life.

3.4 Vent Fluid Circulation

Imagine a deep-sea chimney, spewing superheated water rich in minerals. That's a hydrothermal vent, a window into Earth's fiery heart. But how does the water, heated by magma miles below, actually circulate through this complex system? Understanding vent fluid circulation is key to grasping the entire vent ecosystem.

The process begins deep within the Earth's crust. Hot, chemically-rich fluids, heated far beyond the boiling point due to the immense pressure, percolate through fractured basalts. These fractured zones act as permeable pathways, allowing the fluids to move upwards. The exact pathways are complex, influenced by factors like the permeability of the rock, the temperature and pressure gradients, and the presence of any existing fractures and fissures. It's not a simple straight-up path; it's more of a three-dimensional network of interconnected channels.

As the hot fluid rises, it interacts with the surrounding cold seawater. This interaction causes significant chemical changes. Minerals precipitate out of solution as the fluid cools, creating the characteristic chimney structures we observe. These chimneys aren't just passive structures, though; their porosity and permeability also play a role in controlling fluid flow. The intricate structure of these chimneys, with their layered formations and internal channels, affects the speed and direction of the circulating fluid.

The interplay between the hot, buoyant vent fluid and the cooler, denser surrounding seawater creates convection currents. These currents are crucial for maintaining the circulation, driving the fluid upward and then outwards, spreading the heat and chemicals. The rate of this circulation depends on several factors including the temperature difference between the vent fluid and the surrounding seawater, the permeability of the surrounding rocks and the shape and size of the vent itself. Scientists use various techniques to study and model these dynamic processes, including sophisticated computer simulations and direct observation using submersibles.

Think of it like a giant, underwater geyser. The pressure builds up underground, forcing the superheated fluid upwards through cracks and fissures. The fluid then erupts from the vent, releasing heat and chemicals into the surrounding ocean water. However, the circulation is continuous and more complex than a single eruption. It's a constant cycle of heating, rising, cooling, and sinking, creating a complex network of flows around and within the vent structure itself.

This circulation isn't just important for the physical structure of the vent; it's also essential for the unique ecosystems that thrive around these vents. The chemicals in the vent fluids support chemosynthetic organisms – bacteria and archaea that form the base of the food web. These organisms don't rely on sunlight; instead, they utilize the chemical energy from the vent fluids to produce organic matter. The circulation of the fluid transports these essential chemicals, providing the lifeblood for this unusual ecosystem.

Studying vent fluid circulation helps us understand not only the formation of these remarkable ecosystems but also the larger geological processes occurring within the Earth's crust. The rate and pattern of fluid flow can provide insights into the underlying magmatic activity and the overall hydrothermal system. It's a fascinating interplay of geological and biological processes that demonstrates the interconnectedness of our planet's systems.

Precise mapping of the ocean floor around hydrothermal vent fields using sonar and other advanced techniques is crucial to visualize and analyze this circulation. Careful analysis of fluid samples, collected by specialized equipment deployed from remotely operated vehicles (ROVs) and autonomous underwater vehicles (AUVs), provides valuable data on the chemical composition and temperature of the vent fluids. By combining these observations with sophisticated computer models, scientists can build a comprehensive understanding of the intricate processes driving this circulation.

Furthermore, understanding the dynamics of vent fluid circulation is important for predicting future changes within these systems. Factors like climate change and seafloor spreading rates can influence the temperature and chemistry of vent fluids, affecting the vent's lifespan and the ecosystems they support. Monitoring these vents can provide crucial insights into the Earth's dynamic systems and the delicate balance of marine environments. Moreover, research on vent systems is fundamental in advancing knowledge of extreme environments and their potential for harboring unique life forms. The future of our understanding of these unique systems heavily relies on technological advancements in deep-sea exploration and analysis.

3.5 Vent Ecosystem Dynamics

The vibrant ecosystem surrounding hydrothermal vents is a testament to life's adaptability. These seemingly inhospitable environments, born from the Earth's fiery heart, support a unique community thriving on chemosynthesis rather than sunlight. The vents themselves, spewing superheated, mineral-rich fluids, are the lifeblood of this extraordinary system. Their chemical composition, a mixture of dissolved metals, sulfides, and other elements, fuels the base of the food chain.

Bacteria, archaea, and other microscopic organisms, are the primary producers in this deep-sea oasis. These extremophiles, uniquely adapted to the extreme pressure, temperature, and chemical conditions, harness the energy from chemical reactions to create organic matter. This process, chemosynthesis, forms the foundation of the entire food web. It's a far cry from the sunlit photosynthetic ecosystems we're more familiar with on land.

Larger organisms, such as tube worms, mussels, and clams, rely on symbiotic relationships with these chemosynthetic bacteria. These symbiotic partners provide the energy the larger organisms need to survive. Tube worms, for example, house vast colonies of bacteria within their bodies. These bacteria process the vent fluids, producing energy-rich compounds that are used by the worms for sustenance. This intricate partnership highlights the profound interdependence within the vent ecosystem.

The distribution of life around hydrothermal vents isn't uniform. The immediate vicinity of the vent, with its extreme conditions, is inhabited by organisms uniquely tolerant to these extremes. As you move further away from the vent, the temperature and chemical composition of the water gradually change, supporting a more diverse array of species. This creates a gradient of life, from extremophiles thriving near the source to more familiar deep-sea creatures inhabiting the surrounding areas.

The dynamic nature of the vents themselves impacts the ecosystem. Vent eruptions, though infrequent, can significantly alter the local environment. These events can bury or destroy existing habitats, leading to population shifts and recolonization processes. The fluctuating flow of vent fluids also affects the distribution of resources and, therefore, the density and composition of the surrounding community. Therefore, the vent ecosystem is far from static; it's a constantly evolving system shaped by both biological and geological forces.

Studying these ecosystems provides crucial insights into life's adaptability and the potential for life in extreme environments beyond Earth. The extremophiles residing in these deep-sea oases offer clues to the potential for life on other planets, such as Europa or Enceladus, which also possess hydrothermal systems beneath their icy surfaces. Understanding the intricate connections and interactions within vent ecosystems also helps us comprehend the resilience and vulnerability of deep-sea habitats in the face of environmental changes.

The long-term monitoring of vent ecosystems is becoming increasingly important. Human activities, such as deep-sea mining, pose significant threats to these fragile environments. Understanding the ecosystem's dynamics and the potential impact of human interference is crucial for developing responsible strategies for deep-sea exploration and resource management. The need for conservation and responsible use of these environments is therefore paramount. Protecting these unique ecosystems not only safeguards the biodiversity of the deep ocean but also preserves a valuable window into the origins and limits of life itself.

Furthermore, the geological context of hydrothermal vents is crucial to understanding the ecosystem dynamics. The tectonic plates and their interactions influence the formation and longevity of vents. The composition of the underlying rock layers affects the chemical composition of the vent fluids, thus directly impacting the microbial communities that form the foundation of the food web. Consequently, geological changes in the region may directly alter the life supporting capacity of a vent ecosystem.

The study of vent ecosystem dynamics involves many scientific disciplines. Biologists, geologists, chemists, and oceanographers all collaborate to unravel the complex interactions within these unique environments. Advanced technologies, such as remotely operated vehicles (ROVs) and autonomous underwater vehicles (AUVs), allow scientists to observe and collect data from these inaccessible habitats. This interdisciplinary approach is essential for a holistic understanding of the vent ecosystem's processes and their intricate interconnectedness.

Considering the unique challenges presented by this environment, ongoing research on vent ecosystems is crucial. This includes improved long-term monitoring techniques, enhanced understanding of the specific roles of different organisms within the food web, and improved prediction of geological processes affecting vent activity. This will allow for improved conservation efforts and responsible future exploration of these fascinating deep-sea oases. The more we learn, the better equipped we are to protect these invaluable ecosystems for future generations. Their study offers vital clues not only into the intricacies of deep-sea life, but also into the broader context of life's resilience and adaptability on our planet and potentially beyond.

4 Basalt Petrology

Imagine diving deep, far beneath the waves, to a place where the Earth's crust is raw and new. That's where we find basalt, the dominant rock type of the ocean floor, a testament to the planet's ever-shifting tectonic plates. Basalt, formed from rapidly cooling magma, is a dark, fine-grained volcanic rock, often containing tiny crystals of plagioclase feldspar and pyroxene – its very essence revealing the fiery origins beneath the sea.

Mid-ocean ridges, those underwater mountain ranges where tectonic plates pull apart, are basalt's primary birthplace. Here, molten rock rises from the Earth's mantle, erupting onto the ocean floor and solidifying into sheets of basalt, constantly creating new oceanic crust. This process, known as seafloor spreading, is a fundamental aspect of plate tectonics, constantly reshaping our planet. The composition of mid-ocean ridge basalts (MORB), surprisingly consistent across vast distances, provides valuable clues about the Earth's mantle composition and the processes occurring deep within our planet. Variations do exist, of course, providing a rich tapestry of geological information. These variations, often subtle, can be linked to local variations in mantle source composition, degree of partial melting, and magma mixing.

But the story of basalt doesn't end with its formation. Once formed, these rocks undergo a fascinating metamorphosis. Hydrothermal alteration, a process driven by the hot, chemically-rich fluids circulating through the crust, dramatically changes the basalt's composition and mineralogy. These fluids, heated by magma deep below, rise through cracks and fissures, reacting with the surrounding rocks, often resulting in the formation of new minerals like clays and zeolites. Imagine the interaction of seawater and the intensely hot rock, minerals dissolving and precipitating in an intricate dance.

Hydrothermal vents, themselves fueled by these fluids, are vibrant ecosystems built upon this altered basalt. Their unique biological communities thrive on the chemicals dissolved in the vent fluids. The alteration of basalt isn't merely a chemical change; it's a crucial factor in the formation and maintenance of these extraordinary undersea oases. The textures and colors of altered basalt are telling signs of this process, helping geologists piece together the history of hydrothermal activity in a given area. The degree of alteration can help us understand the intensity and duration of fluid flow, providing a glimpse into the hidden depths.

Weathering, the slow but relentless action of seawater and organisms, also plays a role in shaping the basalt. This process, slower than hydrothermal alteration, involves the breakdown of minerals and the formation of new ones, further altering the rock's physical and chemical properties. Think of how the relentless waves, carrying dissolved salts, slowly break down the exposed rock. Over millennia, the combined effects of hydrothermal alteration and weathering transform the basalts, creating a rich diversity of rocks with unique stories to tell. Each layer, each change, bears witness to an underwater history played out over unimaginable timescales.

Studying basalt petrology, therefore, isn't simply about identifying rocks; it's about deciphering a planet's history, uncovering the processes that shape our world. By examining the minerals within the basalt, their textures, and the chemical fingerprints left by alteration, scientists can reconstruct the conditions under which the rock formed and the subsequent changes it has undergone. It's like reading an ancient, complex text, where each mineral grain, each subtle chemical variation, forms a sentence in the Earth's narrative. The work is meticulous, requiring expertise in both field observations and laboratory analyses. This analysis goes beyond simple descriptions and uses sophisticated geochemical tools that provide incredible detail.

This understanding of basalt is far more than an academic exercise. It provides crucial insights into plate tectonics, the formation of hydrothermal vents, and the overall evolution of our planet. Moreover, as we delve deeper into the ocean's depths, seeking out mineral resources and understanding the complex ecosystems that thrive there, our knowledge of basalt and its alteration becomes increasingly critical. It informs the search for commercially viable deposits and helps us assess the impact of human activities in the deep ocean. The more we understand basalt and its role, the better equipped we will be to navigate this complex and critical environment sustainably. The future of ocean exploration hinges on our understanding of such seemingly simple stones. After all, the deepest secrets often reside in the most unassuming of places.

4.1 Mid-Ocean Ridge Basalts

Mid-ocean ridge basalts (MORB) are the most abundant volcanic rocks on Earth, forming the foundation of the oceanic crust. They're born from the upwelling magma at divergent plate boundaries, where tectonic plates pull apart, allowing molten rock to rise and erupt. This process, known as seafloor spreading, constantly creates new oceanic crust, pushing older crust away from the ridge. Think of it like a giant conveyor belt of rock, slowly moving away from the mid-ocean ridge. The composition of MORB is remarkably consistent across the globe, making it a key focus for geologists studying plate tectonics and Earth's internal processes. We'll explore what makes these rocks so special and what we can learn from them.

These basalts are predominantly tholeiitic, meaning they're relatively low in silica and rich in iron and magnesium. This reflects their origin in the Earth's mantle, a region rich in these elements. The magma doesn't spend much time in chambers, allowing for minimal crystal fractionation, that means that the minerals forming within the magma don't have time to fully separate. As a result, MORB displays a relatively simple mineral assemblage, often consisting primarily of plagioclase feldspar, clinopyroxene, and olivine. Analyzing the exact proportions of these minerals helps us understand the pressure and temperature conditions under which the magma formed, providing crucial insights into the Earth's mantle dynamics. We can essentially use these rocks as a window into the Earth's inner workings.

The chemical composition of MORB isn't uniform; subtle variations occur due to several factors. For instance, the degree of partial melting of the mantle source plays a significant role. A higher degree of melting yields a more depleted source region and produces a basalt with a different trace element signature, meaning different concentrations of rare elements. Likewise, the magma's ascent rate impacts its crystallization and ultimately, the final composition. A faster ascent means less time for minerals to crystallize and separate from the melt, whereas a slower ascent allows for more extensive fractional crystallization, which can alter the composition quite significantly. Understanding these variations helps us map out variations in mantle composition and dynamics along the mid-ocean ridge systems.

Furthermore, the interaction of MORB with seawater is crucial to understanding its alteration. As the hot basalts erupt onto the seafloor, they react with the cold, chemically active seawater, undergoing hydrothermal alteration. This process changes the original mineralogy and chemical composition of the basalt. Hydrothermal fluids circulate through the porous basalt, dissolving certain minerals and precipitating others, resulting in the formation of secondary minerals like clay minerals and zeolites. The alteration profoundly impacts the physical properties of the MORB, making it more brittle and changing its porosity and permeability. This process is crucial for several aspects of marine geology, from forming hydrothermal vent systems to impacting fluid flow in the oceanic crust, which can then affect seafloor spreading processes.

Studying MORB offers a window into the Earth's interior. By analyzing the chemical and isotopic compositions of these rocks, scientists can deduce information about the mantle source regions, the degree of melting, the processes of magma genesis and evolution, and the processes of seafloor spreading. Isotope geochemistry, in particular, helps reveal the origin and history of the mantle, tracing the movement of plumes and convection cells deep within the Earth. Moreover, through understanding the alteration processes, we gain valuable insights into the cycling of elements between the ocean and the Earth's interior, an intricate process affecting both the composition of seawater and the Earth itself. The vast abundance of MORB makes it a unique and readily available material for this research.

Finally, remember the impact of these discoveries. The exploration and understanding of MORB has far-reaching implications for various fields. It deepens our comprehension of plate tectonics, providing crucial evidence for the theory. This understanding, in turn, aids in predicting earthquakes and volcanic eruptions along mid-ocean ridges and surrounding areas. Moreover, the study of hydrothermal alteration informs us on the potential of seafloor resources, such as polymetallic sulfides and other valuable minerals, concentrated along mid-ocean ridges and formed by hydrothermal activity. By investigating MORB, we not only delve into the depths of the Earth but also explore potential resources and learn to better understand our planet and its processes. The continued study of MORB therefore remains a central focus in oceanographic and geological research, promising further insights into the intricacies of our dynamic planet.

4.2 Basalt Alteration

The ocean floor, a vast and mysterious realm, holds secrets within its basaltic heart. These dark, volcanic rocks, formed at mid-ocean ridges, aren't static; they're dynamic participants in a complex chemical dance with seawater. This interaction, known as basalt alteration, significantly modifies the rock's mineralogy, texture, and chemical composition, leaving behind a fascinating record of past ocean conditions and hydrothermal activity.

Hydrothermal alteration, a primary process, occurs when hot, chemically enriched water circulates through fractured basalt. This heated water, often originating from nearby magma chambers, leaches out certain elements like calcium and magnesium from the basalt, while depositing others like iron, silica, and various metallic sulfides. The resulting changes are dramatic, transforming fresh, glassy basalt into a complex mosaic of altered minerals. Clay minerals, like smectite and chlorite, commonly form, giving the rock a softer, less dense texture. Think of it like a slow, chemical metamorphosis, a subtle shift from one state to another. The extent of alteration often depends on the temperature and the chemistry of the interacting fluids and the duration of the interaction.

Imagine a microscopic landscape within the basalt, a world of shifting minerals and migrating elements. The alteration process isn't uniform; some areas might show extensive alteration, while others remain relatively pristine. These variations offer clues to the pathways of hydrothermal fluids, the temperature gradients within the rock, and the overall history of alteration events. Studying these patterns allows us to reconstruct past hydrothermal systems, understanding their intensity, duration, and evolution over time. It's like deciphering an ancient inscription, each mineral grain hinting at a moment in geological history. It's a process requiring meticulous observation and analysis.

Weathering, another significant alteration process, takes place at the seafloor's surface. Seawater, a weak but persistent chemical agent, interacts with exposed basalt, gradually dissolving minerals and altering the rock's composition. This process, slower than hydrothermal alteration, contributes to the overall chemical evolution of the basalt. The combination of hydrothermal and weathering alteration profoundly impacts the physical and chemical properties of the rock, making it porous and reactive, and altering its density, strength and magnetic susceptibility. The degree of alteration plays a critical role in various geological processes, impacting factors such as porosity and permeability, which are essential for fluid flow and storage within the oceanic crust.

The study of altered basalt provides valuable insights into various aspects of oceanographic science. For example, the chemical composition of altered basalt reveals information about the past chemical evolution of seawater, helping to establish baseline measurements. By analysing the alteration mineral assemblages, we can infer temperature and pressure conditions during the alteration process. This enables geologists to reconstruct the thermal regime of the oceanic crust at different times and locations. Altered basalt also plays an important role in various geochemical cycles, acting as a significant reservoir for many elements and influencing the overall distribution of elements in the Earth's system. This process also has broader implications for understanding global biogeochemical cycles. The extent of alteration can influence the biological communities living within and on the rocks.

Moreover, the study of altered basalt has practical applications. Hydrothermal alteration often concentrates valuable minerals and metals within the rocks, which has significant implications for exploration and potential extraction. Understanding the alteration processes is crucial for the sustainable exploration and development of these resources. Moreover, basalt alteration can be used as a proxy for understanding tectonic processes such as plate movement and seafloor spreading rates. The spatial distribution and patterns of altered basalt can indicate the location, geometry and age of hydrothermal systems. These insights are crucial for understanding the formation and evolution of the ocean floor and the processes shaping our planet. It's not just about rocks; it's about understanding Earth's dynamic systems.

5 Sedimentary Processes

Imagine the ocean floor, not as a flat, featureless expanse, but as a vast, ever-changing tapestry woven from countless tiny particles. These particles, the building blocks of the ocean's sedimentary record, tell a story – a story of life, of climate, and of the powerful forces that shape our planet. Understanding these sedimentary processes is crucial to unraveling the mysteries hidden beneath the waves.

Pelagic sediments, a blanket covering much of the abyssal plains, are formed from the slow, gentle rain of particles from above. Microscopic organisms, drifting through the sunlit waters, eventually sink to the bottom, their skeletal remains accumulating over millennia. Volcanic ash, carried by the wind, adds another layer to this sediment blanket, while cosmic dust, a testament to our solar system's history, provides a subtle yet significant contribution. The composition of these pelagic sediments reflects the conditions in the overlying water column—a treasure trove of information for paleoclimatologists and oceanographers. A careful study of their layers, their chemical make-up, and the tiny fossils they contain can reveal past ocean temperatures, nutrient levels, and even the patterns of ancient ocean currents. The rate of sediment accumulation also reveals significant information, some areas receiving millimeters of sediment a century while others may accumulate much more rapidly.

Turbidites, by contrast, represent a dramatic departure from the quiet accumulation of pelagic sediments. These massive deposits are the result of underwater landslides, powerful currents that race down the continental slopes, carrying with them a chaotic mixture of sand, silt, and mud. Imagine a monstrous underwater avalanche, a sudden torrent of sediment cascading down the steep slopes. This rapid deposition creates unique sedimentary structures, layers that record the intensity and energy of these events. The thickness and extent of turbidite deposits provide clues about the frequency and magnitude of past submarine earthquakes and their influence on the ocean floor's topography. Examining the layers of these formations, scientists can reconstruct the history of the submarine landslides in a particular location, and the impact they had on the surrounding environments and organisms. This is achieved by detailed lithological analysis, which is essential to the reconstruction of events.

The study of sedimentary structures goes far beyond simply identifying the types of sediments present. These structures, such as bedding planes, ripple marks, and graded bedding, offer crucial insights into the environmental conditions under which the sediments were deposited. Bedding planes, for example, represent the boundaries between different sedimentary layers, potentially indicating changes in depositional environments over time. Ripple marks, small-scale undulations formed by currents, can indicate the direction and strength of ancient currents. Graded bedding, where sediment size decreases upwards within a layer, tells the story of a sudden decrease in current velocity as a turbulent current calms down. These structures paint a vivid picture of the past ocean environment, revealing details such as water depth, current speed, and the presence of organisms that might otherwise be undetected. In understanding these features, the sedimentary history of the region becomes more precise.

Combining the study of pelagic sediments, turbidites, and sedimentary structures paints a remarkably detailed picture of the ocean floor's dynamic history. It reveals not only the physical processes that have shaped the ocean basin but also the biological and chemical interactions that have taken place within it. By carefully examining these records, we gain a profound understanding of the Earth's past, its ever-evolving landscapes, and the rich tapestry of life that flourishes within its depths. This knowledge is not merely academic; it is vital for understanding the present and predicting the future, particularly as human activities increasingly impact the delicate balance of the marine environment. It's a fascinating world, this underwater world of sediments. And it is a field where new discoveries are always around the corner. Each grain of sediment holds a secret, waiting to be uncovered.

5.1 Pelagic Sediments

Pelagic sediments, a seemingly simple term, encompass a vast and fascinating world of tiny particles that slowly accumulate on the deep ocean floor. Imagine a constant, gentle rain of microscopic shells, dust from continents carried on the wind, and volcanic ash settling over millennia. This continuous deposition creates layers, a historical record of the ocean's past, revealing changes in climate, biological productivity, and even volcanic activity. The composition of these sediments, often dominated by the skeletal remains of microscopic organisms like diatoms and foraminifera, acts as a proxy for past ocean conditions. Analyzing their chemical makeup can tell us about ancient ocean temperatures, salinity levels, and the abundance of nutrients in the water column. For instance, the presence of certain types of foraminifera can indicate past ice ages, while variations in the isotopic ratios of oxygen within their shells provide further clues.

The study of pelagic sediments isn't solely about the past; it also provides insights into present-day ocean processes. For example, the rate of sediment accumulation can reflect current biological productivity, offering clues about the health of the marine ecosystem. Furthermore, the distribution patterns of these sediments can illuminate ocean currents and their influence on the transport of particles across vast distances. Consider how the distribution of siliceous sediments—those rich in the silica shells of diatoms and radiolarians—often aligns with regions of high productivity, demonstrating the close relationship between biological activity and sediment formation. Changes in these distribution patterns over time can therefore signal shifts in ocean circulation or nutrient availability.

Different types of pelagic sediments tell distinct stories. Calcareous sediments, composed mainly of calcium carbonate from the shells of organisms like coccolithophores and foraminifera, are abundant in warmer, shallower regions of the ocean. Their presence or absence in certain areas can reflect the depth of the carbonate compensation depth (CCD), a critical boundary where the rate of calcium carbonate dissolution exceeds its rate of deposition. This depth is influenced by factors such as ocean temperature, pressure, and the concentration of carbon dioxide in the water. Studying changes in the CCD through time embedded in the sediment layers provides a record of how the ocean's chemistry and environment have changed.

The analysis of pelagic sediments relies heavily on careful sampling and laboratory techniques. Scientists use specialized equipment to collect sediment cores, cylindrical samples that provide a vertical profile through the sediment layers. These cores are then meticulously analyzed in laboratories, with techniques ranging from microscopic examination to geochemical analysis to unravel the complex histories locked within. Careful measurements of the grain size, mineral composition, and chemical isotopes provide a wealth of data that can be used to reconstruct past ocean environments and understand how these environments have evolved over time. The layers of sediment are chronological records. Radioactive dating techniques can be employed to determine the age of specific layers, providing a precise timeline of events.

Furthermore, the study of pelagic sediments is crucial for understanding the long-term impacts of human activities on the oceans. The accumulation of pollutants and microplastics in these sediments serves as a clear indicator of the extent of human influence on marine ecosystems. By studying the composition and concentration of these contaminants in sediment cores, scientists can assess the timing and extent of pollution events and predict their long-term consequences. This information is invaluable for developing effective strategies for mitigating pollution and protecting the health of the oceans. The ongoing monitoring and analysis of pelagic sediments are therefore vital for understanding the present and future state of our planet's oceans, providing a detailed chronicle of its biological, chemical, and physical evolution over millions of years. Studying these sediments gives us a window into how the oceans change and how those changes could impact the planet, a story etched in the fine particles on the ocean floor, layer upon layer, year upon year. The ongoing research on this delicate deep sea archive is vital for understanding how our planet works and is constantly changing and how humans are impacting these changes.

5.2 Turbidites and their Formation

Imagine a colossal underwater landslide, a churning avalanche of sediment hurtling down a continental slope. This isn't a gentle drift; it's a catastrophic event, a powerful current carrying tons of sand, silt, and clay. This immense, chaotic flow, known as a turbidity current, dramatically alters the deep-sea floor, leaving behind distinctive rock formations we call turbidites. These aren't just random piles of sediment; they hold clues to past environmental conditions, tectonic activity, and the dynamic processes shaping our planet's oceans.

Turbidites form in specific environments, typically on the continental slopes and rises, where the steep incline facilitates rapid sediment transport. The initial trigger can vary; an earthquake shaking loose vast amounts of sediment, a submarine slump destabilizing the slope, or even a large storm surge can initiate this destructive yet fascinating process. The sediment, initially suspended in the water column, forms a dense, gravity-driven flow that cascades downslope, eroding and incorporating more material along its path.

The energy of the turbidity current determines the size and characteristics of the resulting deposit. Powerful currents can transport coarse grains like sand and gravel for considerable distances, while weaker currents may only carry finer materials like silt and clay. This graded bedding, a hallmark of turbidites, is one of the key features geologists look for when identifying these formations. The largest grains settle first, followed by progressively finer grains as the current slows down; imagine a carefully layered sediment cake baked by the power of nature.

The speed and volume of the flow also influence the thickness and extent of the turbidite deposits. A single catastrophic event can result in a thick layer of sediment, while repeated smaller events may create thinner, interbedded layers. The resulting structure is often a complex interplay of graded beds, exhibiting Bouma sequences—a classic pattern of layering that reflects the gradual deceleration of the turbidity current. These sequences, characterized by distinct divisions (Ta, Tb, Tc, Td, Te), offer insights into the dynamics of the depositional process, providing a window into the past behavior of these formidable undersea flows.

The composition of the turbidite sediment is crucial for understanding its source and the environmental context of its formation. The sediment itself can contain a mixture of continental and marine debris, including fragments of rocks, fossils, and minerals. Analyzing this sediment composition—the proportions of quartz, feldspar, and other minerals—provides valuable information on the source region and the processes that affected the sediment before and during transport. The presence of specific fossils can help to date the turbidite layer, allowing scientists to build a chronology of these events.

Mapping and analyzing turbidite sequences is essential for understanding the geological history of ocean basins. These formations act as valuable records of past tectonic activity, sea-level changes, and even catastrophic events. By studying the distribution, thickness, and composition of turbidites, geologists can reconstruct the history of submarine landslides, identify past earthquake events, and obtain insights into the processes that shape the deep ocean environment. This information informs our understanding of hazards, such as tsunami risks and the potential for future slope failures, enhancing the ability to assess and mitigate these geological threats.

But the study of turbidites goes beyond merely understanding past events; they also play a significant role in several key geological applications. For instance, the analysis of turbidite sequences allows for the creation of robust geological models, which aid in the location and assessment of potential hydrocarbon reservoirs. The ability to effectively predict the location and characteristics of turbidite reservoirs is of great economic importance to the oil and gas industry, and these deposits provide valuable clues and insights into such regions. Additionally, the high permeability and porosity of certain turbidites make them potential aquifers, playing a role in the study of submarine groundwater. The properties and characteristics of turbidites are not only key elements for scientists, but also provide crucial insight into the economics of subsurface exploration and resource extraction. Therefore, the exploration of turbidites and their formation holds deep scientific significance, but also carries significant economic implications in various aspects of resource management.

5.3 Sedimentary Structures

Imagine a vast, underwater landscape, a world hidden beneath the waves. This is where we find sedimentary structures, silent witnesses to millions of years of ocean history. These structures, formed by the slow accumulation and movement of sediment, hold clues to past ocean currents, the nature of ancient life, and the very forces that shaped our planet. Studying them is like reading a fascinating, layered book, each page a different era.

Understanding sedimentary structures begins with grasping the different types of sediment. We have fine-grained clays, carried by slow currents, settling gently to form smooth, continuous layers. Then there are coarser sands, deposited by stronger currents, often creating cross-bedding – inclined layers showing the direction of the ancient flow. These patterns aren't random; they reveal the power and direction of the currents that once swept across the ocean floor.

Picture this: a powerful underwater landslide, a turbidity current, a dense rush of sediment and water cascading down a slope. This leaves its mark on the seabed as a distinctive layer called a turbidite. They are often graded bedding, meaning the largest grains settle first, followed by progressively finer particles, creating a layered effect that's easily recognizable to geologists. We can even deduce the speed and volume of the ancient current from the thickness and grain size variation within a single turbidite layer.

Bioturbation, another important process, is the disruption of sediment layers by living organisms. Worms, burrowers, and other creatures constantly move through the sediments, creating traces, trails, and burrows. These traces, even when the organism itself is long gone, provide invaluable information. They tell us about the types of organisms living on the seafloor at that time, the oxygen levels in the water, and the consistency of the sediment. These organisms act as tiny excavators and architects, continuously reshaping the layers below.

Then consider ripple marks, those delicate wave-like patterns often found on sedimentary surfaces. These aren't just pretty patterns; they're clues to the ancient currents' strength and direction. Symmetrical ripple marks indicate oscillating currents, like waves gently lapping at a beach. Asymmetrical ripples, on the other hand, tell a story of unidirectional currents, like a river flowing steadily in one direction. The size and spacing of these ripples give us additional details. Small ripples suggest weaker currents, while large ones indicate stronger flows.

Studying sedimentary structures is more than just identifying layers. It's about piecing together a narrative. By carefully observing the types of structures present—the grain size, layering, bioturbation, ripple marks, and more—we can reconstruct the past environment. We can determine whether the seabed was a tranquil, slowly accumulating plain or a dynamic zone of powerful currents and frequent landslides. We can glimpse the ancient life that thrived there, infer the past oceanographic conditions, and gain a deeper appreciation for the ever-changing nature of the ocean floor.

Furthermore, understanding these structures is crucial for interpreting the geological history of the ocean basin. Sedimentary sequences, with their distinct layers and structures, act as chronological records, preserving evidence of past events. Each layer represents a moment in time, a snapshot of the ocean's past. The careful study of these layers, painstakingly comparing and analyzing their structures, is how we piece together a coherent picture of millions of years of geological activity.

This detailed analysis allows us to understand not only the depositional environment but also the tectonic history of the region. For example, the presence of particular structures can be indicative of past tectonic upheavals, such as earthquakes or volcanic activity. These events, often catastrophic, leave their mark on the sediment layers in the form of distinct deformations and unusual layering patterns. By studying these anomalies, we can decipher the timeline of major geological events that shaped the seabed.

Moreover, sedimentary structures are vital for understanding the distribution of resources on the ocean floor. Many valuable resources, such as hydrocarbons and minerals, are associated with specific sedimentary environments and structures. For instance, certain types of sedimentary basins are known to be favorable locations for the accumulation of oil and gas deposits. By understanding the sedimentary structures within these basins, geologists can better predict the location and extent of these resources. This knowledge is crucial for efficient and responsible exploration and extraction.

Finally, the study of ocean floor sedimentary structures plays a significant role in understanding past climate change. The composition of sediment layers reflects changes in ocean chemistry and biological productivity. By analyzing the properties and structures of these layers, we can obtain valuable information on past climate conditions, including sea level fluctuations and temperature variations. This data provides critical insights for modeling future climate change scenarios and for managing the impact of ongoing climate change. The ocean floor is a vast archive of climate data, ready to be read by those who study these delicate structures. Each layer holds a piece of the puzzle. It is by piecing together these pieces that we can paint a clear picture of the past and predict the future, helping us safeguard this precious environment.

6 Advanced Mapping Techniques

Imagine piloting a tiny submarine, a robotic explorer venturing into the inky blackness of the deep. That's essentially what Autonomous Underwater Vehicles (AUVs) do, silently mapping vast swathes of the ocean floor with unprecedented detail. Equipped with sophisticated sonar systems far exceeding the capabilities of older technologies, these unmanned submersibles gather incredibly precise bathymetric data – creating incredibly detailed maps of the seabed's topography. This allows us to pinpoint previously unseen features, from towering underwater mountains to hidden canyons. The AUV's journey isn't just about mapping; it's about discovering.

Remotely Operated Vehicles (ROVs), on the other hand, offer a different perspective. Tethered to a surface vessel, these robotic workhorses provide real-time visual feedback. This isn't just about pretty pictures; ROVs are equipped with high-resolution cameras, manipulators, and various sensors, allowing scientists to collect samples, perform in-situ experiments, and carefully examine geological features. They can investigate hydrothermal vent systems up close, collecting data on temperature, chemical composition, and biological communities. Think of them as the deep-sea geologists' tireless assistants, their cameras peering into places where human eyes could never reach. With ROVs, we can obtain detailed images of rock formations, allowing for more accurate petrological analysis compared to the indirect methods of the past.

The data collected by AUVs and ROVs is far more detailed than anything we've previously had access to. This allows for a level of detail in our understanding never before achieved. This data becomes the foundation for creating incredibly sophisticated 3D models of the ocean floor. These aren't just pretty visualizations; they're powerful tools for scientific analysis, allowing researchers to understand the complex interplay of geological processes, from plate tectonics and volcanism to sedimentation and erosion. Imagine being able to virtually "fly" through a canyon, examining rock formations from every angle, and understanding the geological history embedded within those ancient layers. This is the power of 3D ocean floor modeling. These models are essential not just for academic research but also for practical applications, such as identifying potential mineral resources, assessing the stability of underwater infrastructure, and managing marine ecosystems. Detailed knowledge of the seafloor's topography and subsurface structure is crucial for effective management of this vital resource.

Creating these 3D models is a multi-stage process. First, the raw data needs careful processing and cleaning. This involves removing noise from the sonar data and correcting for various environmental factors. Then, the processed data is integrated to create a high-resolution digital elevation model (DEM). Specialized software then uses this DEM to construct the 3D model, often incorporating data from other sources, such as magnetic surveys and seismic reflection profiles, to build a complete picture. The final model is an interactive, three-dimensional representation of the ocean floor, allowing scientists to examine geological structures from any angle. Further analysis might involve integrating geological interpretation, adding information on rock types and geological ages.

The integration of these advanced mapping techniques with other data sources offers a transformative perspective on our understanding of the ocean floor. By combining bathymetric data with information about water chemistry, temperature gradients, and biological communities, scientists can create even more comprehensive models of the deep ocean. These integrated models can then be used to understand the interconnectedness of geological, chemical, and biological processes within the ocean's environment, providing an entirely more holistic understanding of the deep ocean's delicate dynamics. This understanding is fundamental to our efforts to protect this unique environment and understand the complex interplay of the deep sea with the global environment. The advanced mapping techniques are not merely technological improvements; they represent a shift in how we view and interact with the planet. They open up a world of potential discoveries, expanding our knowledge of Earth's processes and uncovering new challenges and opportunities, requiring new methodologies of scientific exploration and improved data analysis strategies. The ability to create highly-accurate, interactive 3D models of the ocean floor is paving the way for a deeper understanding of our planet.

6.1 AUV and ROV Surveys

Autonomous Underwater Vehicles (AUVs) and Remotely Operated Vehicles (ROVs) are revolutionizing how we explore the ocean depths. These sophisticated tools allow for detailed surveys of the seafloor, far surpassing the capabilities of traditional methods. Imagine a small, robotic submarine, equipped with an array of sensors, silently gliding through the inky blackness, collecting data previously inaccessible to humans. That's the power of AUVs. They can explore vast areas autonomously, mapping the terrain, taking samples, and collecting valuable scientific information. Their endurance and ability to cover vast distances make them ideal for large-scale mapping projects.

Unlike AUVs, which operate independently, ROVs are tethered to a surface vessel, providing a real-time connection to operators. This allows for more interactive exploration, with operators able to direct the ROV's movements and manipulate its tools remotely. Imagine the excitement of controlling a sophisticated machine thousands of feet beneath the waves, guiding it to inspect a hydrothermal vent or analyze a unique rock formation. ROVs, with their maneuverability and immediate feedback, are essential for detailed inspections and targeted sample collection. They also provide a higher level of precision, allowing scientists to pinpoint locations of interest with incredible accuracy.

The data collected by both AUVs and ROVs is invaluable. High-resolution sonar images create detailed maps of the seafloor, revealing intricate details of its topography. Cameras capture stunning images and videos, allowing scientists to witness the wonders of the deep-sea environment firsthand, from strange lifeforms to unique geological formations. Specialized sensors measure water temperature, salinity, and chemical composition, providing valuable insights into the ocean's complex systems. Sophisticated robotic arms enable the collection of sediment and rock samples for detailed laboratory analysis. These samples contribute significantly to our understanding of oceanic processes.

Combining AUV and ROV data creates a comprehensive picture of the ocean floor. AUVs provide broad-scale mapping, identifying areas of interest, while ROVs conduct detailed investigations of specific sites. The synergistic use of these technologies allows for a more efficient and comprehensive understanding of the seabed. This combined approach is particularly important for studies on submarine volcanoes, hydrothermal vents, and the distribution of seafloor resources. Integrating data from both systems also helps in generating 3D models of the seafloor, allowing scientists to visualize complex geological features with unprecedented clarity and accuracy.

One exciting application is the study of hydrothermal vents. These underwater geothermal springs support unique ecosystems, and AUVs and ROVs are critical in studying their formation, chemical composition, and biological communities. AUVs can map the vent fields extensively, determining their spatial distribution and size. ROVs can then perform close-range observations, collecting samples of vent fluids and biological specimens. This detailed information helps scientists understand the chemical reactions occurring within the vents, the biological adaptations of life in these extreme environments, and the overall dynamics of these unique ecosystems.

The integration of AUV and ROV technologies is also essential for effective resource exploration. These underwater robots can be used to map the distribution of mineral deposits, gas hydrates, and other valuable resources on the seafloor. The detailed data they collect helps to identify areas with high potential for resource extraction, aiding in the responsible and sustainable development of these resources. By using these technologies and performing detailed surveys, we can better estimate the quantity, quality, and accessibility of these deep-sea resources.

However, despite these advancements, challenges remain. The immense pressure, darkness, and cold of the deep ocean present significant technical hurdles. The cost of AUV and ROV operations can also be substantial, particularly for long-duration missions. Data processing and analysis can also be complex, demanding powerful computers and specialized software to interpret and visualize the massive amounts of data collected. Nevertheless, ongoing technological advancements are continually improving the capabilities and affordability of these tools, allowing for even more extensive exploration of our planet's largely unexplored ocean floor. The future of ocean exploration, and our understanding of this vast and mysterious environment, hinges on continued innovation and application of these amazing technologies. The data collected by AUVs and ROVs will undoubtedly lead to new discoveries and a deeper appreciation of our planet's incredible diversity.

6.2 3D Ocean Floor Modeling

Creating a three-dimensional model of the ocean floor isn't just about pretty pictures; it's about unlocking a wealth of scientific understanding. Imagine being able to visualize the intricate details of a submerged volcano, the precise path of an underwater landslide, or the complex network of hydrothermal vents. This level of detail is achievable through 3D modeling, a powerful tool that transforms raw data into insightful, interactive representations.

The process begins with data acquisition. Various techniques, which you've already learned about—sonar, seismic reflection, and even satellite altimetry—provide the foundational information.

Sonar creates a detailed bathymetric map, essentially a contour map of the seafloor, showing its ups and downs. Seismic reflection adds another dimension by revealing the subsurface layers, like looking at a geological cross-section beneath the ocean's surface. These datasets, though incredibly valuable on their own, come to life when combined within a 3D framework.

Different software packages are available, each offering a unique set of capabilities and functionalities. Some are designed for simple visualization, allowing you to rotate and zoom in on the model, highlighting specific features. Others provide more sophisticated tools for analysis, enabling you to measure volumes, calculate slopes, and even model fluid flow through porous rocks—crucial for understanding hydrothermal vent systems. Learning to use these programs requires practice and patience, but the results are worth the effort. Think of it like learning a new language—initially challenging, but incredibly rewarding once mastered.

One of the most exciting applications of 3D ocean floor modeling is in geological hazard assessment. By creating realistic models of areas prone to underwater landslides or volcanic eruptions, scientists can better predict the potential impact of these events. This information is invaluable for coastal communities, helping them prepare for potential tsunamis or other hazards. Imagine the lives saved by having this level of foresight. It changes the response from reactive to proactive.

Beyond hazard assessment, 3D modeling plays a critical role in exploring for valuable resources. Oil and gas exploration often relies on detailed subsurface imaging, providing clues to the location of potential reservoirs. Similarly, the search for polymetallic nodules, rich in cobalt and other valuable metals, necessitates precise mapping of the seafloor. These resources are essential for the global economy, and 3D modeling significantly improves the efficiency and effectiveness of their exploration. Think of how these models guide underwater robotic vehicles, enabling them to pinpoint the location of resources.

Furthermore, 3D modeling significantly enhances our understanding of geological processes. By visualizing the complex interactions between tectonic plates, volcanic activity, and sediment transport, scientists can gain a deeper understanding of how the ocean floor evolves over time. These insights aren't merely academic; they are crucial for predicting future changes and managing the impacts of human activity. Think about the long-term implications of understanding how ocean currents impact sediment distribution, or how volcanic eruptions shape the underwater landscape. The models show you those interactions in a way that's impossible with any other method.

Ultimately, the power of 3D ocean floor modeling lies in its ability to synthesize vast amounts of data into a comprehensible, interactive format. It transforms abstract concepts into tangible visualizations, making complex geological processes accessible and engaging. This is invaluable, not just for experienced scientists but also for those new to the field, providing a powerful tool for learning and exploration. The technology continues to evolve, promising even more detailed and sophisticated models in the future, further enhancing our understanding of our planet's hidden depths. By mastering this technique, you equip yourself with a key tool in unlocking the secrets of the deep ocean.

7 Future of Ocean Exploration

The ocean, a vast and mysterious realm, holds secrets yet to be uncovered. Technological advancements are poised to revolutionize how we explore it, pushing the boundaries of our understanding far beyond current capabilities. Imagine autonomous underwater vehicles (AUVs) equipped with advanced sensors capable of mapping the ocean floor with unprecedented detail, revealing hidden hydrothermal vents and previously unknown geological formations. These robotic explorers, working in concert with remotely operated vehicles (ROVs) piloted from the surface, will paint a much more complete picture of the ocean floor's intricate tapestry.

Miniaturization of sensors will allow for greater precision and wider deployment. We can envision swarms of tiny robots, each carrying a specialized sensor, working together to collect data on water chemistry, temperature, and biological activity across vast areas. This data will be processed using advanced algorithms that can detect patterns and anomalies, highlighting areas worthy of further investigation. This shift from large, expensive expeditions to smaller, more targeted operations will make ocean exploration more efficient and cost-effective.

Many fundamental questions remain unanswered. Precisely how do the incredibly diverse ecosystems around hydrothermal vents function? What are the exact mechanisms driving their formation? What other undiscovered life forms lurk in the abyssal plains and hadal trenches? The answers to these and countless other questions could lead to breakthroughs in fields ranging from medicine to materials science. The potential for discovery is simply staggering.

Future exploration strategies will necessitate a multifaceted approach. International collaborations will be essential, bringing together scientists, engineers, and policymakers from across the globe. Shared data platforms will need to be developed to ensure data accessibility and facilitate collaborative analysis. A focus on sustainability is imperative. Our exploration must not damage the very environment we are studying.

The economic implications of ocean exploration are profound. The ocean floor contains vast reserves of valuable minerals and energy resources. Developing sustainable methods for extracting these resources will be crucial, balancing economic benefits with environmental protection. The ocean's potential to provide food and renewable energy sources is also significant, demanding careful stewardship.

Environmental considerations will be paramount. The fragile ecosystems of the deep ocean must be protected from the impacts of exploration and resource extraction. Developing environmentally friendly technologies and implementing strict regulations will be key to minimizing our environmental footprint. Sustainable ocean exploration necessitates a careful balance between progress and conservation, a paradigm shift that prioritizes the long-term health of the ocean's delicate ecosystems. We must remember that the ocean's health is inextricably linked to our own. The future of ocean exploration is not just about discovery; it's about responsible stewardship. Our actions today will determine the ocean's fate for generations to come. The choices we make will shape the narrative of future exploration, ensuring its sustainability and ethical integrity. We stand at the precipice of a new era of discovery. The future holds immense potential, but only if we approach it with both ambition and respect. The deep's delicate delights are worth protecting.

7.1 Technological Advancements

The ocean's depths hold secrets beyond our current grasp, secrets that advanced technology is slowly unveiling. Imagine submersibles, not just the clunky, tethered robots of the past, but sleek, autonomous vehicles capable of navigating the deepest trenches, equipped with sensors far exceeding anything we have today. These autonomous explorers will map previously uncharted territories with unprecedented accuracy, providing a detailed 3D model of the ocean floor, revealing hidden hydrothermal vents, and even identifying previously unknown geological formations.

Next-generation sonar systems, employing advanced signal processing and artificial intelligence, will pierce through the murky depths, penetrating sediment layers to reveal high-resolution images of the sub-seafloor structure. We'll see not just the surface of the seafloor but the intricate network of faults, fissures, and magma chambers beneath. This level of detail will allow for a far more accurate understanding of plate tectonics and volcanic activity along mid-ocean ridges, revealing the processes that shape our planet in ways never before imagined.

Furthermore, developments in remote sensing are already providing us with invaluable data from space. Satellites, equipped with increasingly sensitive instruments, monitor subtle changes in sea level, gravitational anomalies and even ocean temperature variations – all subtle clues hinting at undersea geological activity, both recent and long past. The integration of these various data sources – sonar, seismic reflection, and satellite remote sensing – will create a comprehensive and dynamic picture of the ocean floor, a picture far more complete than anything we currently possess.

Think of the potential implications. We could predict volcanic eruptions with much greater accuracy, potentially mitigating their impact on coastal communities. A deeper understanding of hydrothermal vents might unlock new insights into the origins of life on Earth, as well as revealing potential resources like rare earth minerals which are vital for developing innovative technologies. The ability to model complex oceanographic processes could help predict the impact of climate change on marine ecosystems.

The advancements in materials science are also pivotal. Submersibles require materials capable of withstanding immense pressure, and the development of new lightweight yet incredibly strong alloys will allow us to explore the deepest parts of the ocean more easily and safely. Similarly, advancements in energy storage technology will extend the operational range of these vehicles, enabling longer missions and deeper explorations. These are not merely incremental improvements; they are paradigm shifts.

Improved sensors are crucial. Sophisticated chemical sensors will analyze the composition of hydrothermal vent fluids with greater precision, providing valuable insights into their chemical processes and their role in supporting unique ecosystems. Advanced biological sensors will identify and catalog the life forms inhabiting these extreme environments, increasing our understanding of biodiversity and adaptability. High-resolution cameras will capture breathtaking images of the ocean's depths, allowing the public to share in the wonder of these discoveries.

Data analysis itself will undergo a revolution. The sheer volume of data generated by these advanced technologies demands new techniques for processing and interpreting it all. Machine learning algorithms, capable of identifying patterns and anomalies that might escape human observation, will play a key role in uncovering hidden secrets within the data. This will accelerate the pace of scientific discovery, allowing us to answer questions about the ocean floor far more rapidly than ever before.

Finally, consider the ethical implications. As our ability to explore and exploit the ocean floor increases, it is vital that we do so responsibly. Careful environmental impact assessments must be conducted before any resource extraction takes place. International cooperation is essential to ensure that the ocean's resources are managed sustainably for the benefit of all humankind. The future of ocean exploration is not just about technological advancement, but about responsible stewardship of our planet's most mysterious and valuable realm. The future of this field lies in the hands of young scientists—scientists who are passionate about the ocean and dedicated to its responsible exploration.

7.2 Unanswered Questions

The ocean's mysteries run deeper than any submersible has yet explored. We've mapped vast swathes of the seafloor, identified hydrothermal vents teeming with life, and categorized the dominant rock types. Yet, fundamental questions remain unanswered. For instance, the exact mechanisms driving the long-term evolution of mid-ocean ridge spreading centers are still debated. While we understand the basics of plate tectonics and magma upwelling, the intricate interplay of these forces, leading to variations in spreading rates and magma composition across different ridge segments, continues to challenge our understanding. Precise models accounting for these variations, predicting future spreading patterns, and anticipating potential shifts, are still under development, hindering comprehensive predictive capabilities.

Consider the enigmatic nature of some deep-sea sedimentary layers. Although we can identify many pelagic sediments, the origins of certain unique sedimentary compositions remain puzzling. The source materials, transport mechanisms, and depositional environments for these unusual layers present ongoing challenges for researchers. Further research into the chemical fingerprints of these sediments could unveil crucial information about past ocean conditions, climatic changes, and even extraterrestrial influences. Advanced isotopic analysis coupled with sophisticated modeling techniques could resolve these outstanding questions, furthering our understanding of deep-time oceanographic processes.

The complexities of hydrothermal vent systems offer another realm of unanswered questions. While we know these vents support unique ecosystems, the precise interplay between geological processes, chemical fluxes, and biological adaptations remains to be fully understood. The long-term sustainability of vent ecosystems, their vulnerability to environmental changes, and the full extent of their biodiversity, all pose significant scientific uncertainties. Long-term monitoring, coupled with advanced in-situ experimentation, is required to unlock the secrets of these extraordinary environments. This requires innovative technological advancements, enabling continuous data acquisition over extended periods without significantly impacting the fragile vent ecosystems.

Beyond the biological aspects, the geochemical processes within hydrothermal vents need further investigation. We understand the basic chemistry of vent fluids, but the kinetics of metal precipitation, the formation of unique mineral structures, and the role of microbial communities in these processes all warrant further study. This includes investigating the potential for the formation of unique mineral deposits of economic value, while understanding potential risks related to resource extraction. A multidisciplinary approach, combining geochemical modeling with advanced analytical techniques, could provide a deeper understanding of the chemical dynamics within these complex systems, potentially leading to significant technological advancements and insights into mineral formation.

Finally, the future of ocean exploration itself hinges on the answers to several crucial questions. What technological breakthroughs are needed to reach the deepest, least-explored parts of the ocean floor? How can we balance the need for resource extraction from the ocean floor with the urgent imperative for environmental protection? And, perhaps most importantly, how do we effectively communicate the importance of ocean exploration and conservation to a global audience, ensuring long-term support for sustained research and protection of this critical ecosystem? These questions represent not only scientific challenges but also significant societal and political considerations. The next generation of ocean explorers will need to embrace interdisciplinary approaches and collaborations to address these multifaceted challenges. The answers may lie in a renewed focus on innovative technological solutions, international collaborations, and impactful public outreach.

7.3 Exploration Strategies

Ocean exploration is a thrilling adventure, but effective strategies are crucial for success. Planning isn't just about choosing the right tools; it's about understanding your goals and tailoring your approach. For instance, if your focus is on identifying new hydrothermal vent fields, you'll prioritize techniques like high-resolution sonar mapping and the deployment of AUVs equipped with chemical sensors to detect unique fluid signatures. This targeted approach, in contrast to a broad survey, significantly boosts efficiency.

We must also prioritize collaboration. Oceanography is inherently multidisciplinary. Successfully mapping a region requires geologists, geophysicists, biologists, and engineers all working together. The diverse expertise enables a complete understanding, revealing not only geological structures, but also the ecological impacts and potential economic opportunities associated with a given area. Synergistic efforts ensure a comprehensive data set, much richer than a collection of independent studies. This collaboration is invaluable in identifying correlations between geological features, fluid chemistry, and biological communities within hydrothermal systems.

Data analysis plays a pivotal role. Collecting massive datasets from sonar scans, seismic surveys, and AUV sensors is only the first step. Sophisticated processing and modeling are essential to interpret the data effectively. For example, 3D seismic reflection data, when correctly processed, can reveal subtle variations in subsurface rock layers, potentially indicating the presence of mineral deposits or undiscovered hydrothermal systems. We should also focus on visualizing the data using advanced software capable of producing stunning 3D models which allow us to study the complex interplay between various geological structures and processes in far greater detail.

Consider the environmental impact. Sustainable exploration is paramount, minimizing the disturbance to the delicate ocean ecosystems. This mandates meticulous planning of sampling locations, the careful deployment of equipment, and the implementation of robust environmental monitoring programs, to ensure that the scientific gains don't come at the expense of the very environment under study. Careful adherence to strict protocols guarantees the health and integrity of the ocean floor for future exploration and research. The development and application of environmentally friendly technologies are also vital for mitigating the negative impacts of our investigations.

Funding is a critical aspect, Securing funding for ocean exploration requires a compelling narrative, emphasizing the scientific and societal value of the research. Proposals must effectively communicate the potential for discovery, the anticipated results, and the long-term benefits of both the knowledge gained and the advancement of technologies, thereby ensuring the long-term sustainability of our exploration efforts. Clearly articulated plans that showcase the practical applications of the research, particularly within the realm of resource management and environmental protection, significantly increase the chances of securing the necessary funding. A strong case for a particular area of investigation must also include plans for data dissemination and educational outreach.

Risk assessment is vital. Ocean exploration involves inherent risks, from equipment malfunctions to extreme environmental conditions. Thorough risk assessment, including detailed contingency plans, is essential. This involves evaluating potential hazards like underwater currents, geological instability, and the risks associated with operating sensitive equipment in a harsh environment. A robust risk management strategy, built upon meticulous planning and thorough preparation, ensures not only the success of the expedition, but importantly the safety of the researchers and the preservation of equipment. It is wise to create simulations of various scenarios to refine strategies and better prepare the team.

Finally, remember that exploration is a continuous process. Each expedition builds upon previous knowledge, refining techniques, and prompting new questions. Sharing data openly and fostering international collaboration are crucial for accelerating progress and expanding our collective understanding of the ocean floor. By working together, we can unlock its secrets, discovering valuable resources and gaining a deeper appreciation for the Earth's complex geological history and the delicate ecosystems that thrive in its deepest depths. Our understanding of the ocean continues to evolve, demanding a dynamic and adaptive approach to future exploration efforts.

7.4 Economic Implications

The ocean floor holds immense untapped resources. Polymetallic nodules, rich in manganese, cobalt, nickel, and copper, lie scattered across the abyssal plains. These nodules, slowly accumulating over millions of years, represent a potentially significant source of these crucial metals, vital for modern technologies like batteries and electronics. However, their extraction presents considerable challenges, including the immense depths, the fragility of the deep-sea ecosystem, and the lack of established mining regulations. The economic potential is undeniable, yet sustainable and responsible harvesting remains a critical hurdle.

Developing the technologies for deep-sea mining is an expensive undertaking. Submersibles capable of withstanding the crushing pressures of the deep ocean, remotely operated vehicles (ROVs) equipped with sophisticated tools, and extensive support infrastructure are all necessary. The initial investment is substantial, and the risks are high. Furthermore, the environmental impact needs thorough assessment and mitigation strategies must be implemented to prevent irreversible damage to the fragile deep-sea ecosystems. Careful economic modelling, balancing potential profits with environmental costs, is therefore essential.

Beyond polymetallic nodules, the ocean floor also holds potential for other valuable resources. Hydrothermal vents, while fascinating ecosystems in their own right, contain valuable minerals dissolved in their superheated fluids. These fluids precipitate minerals, forming unique deposits that can be rich in metals like gold, silver, zinc, and lead. However, extracting these resources near active hydrothermal vents poses significant challenges, not just technological, but also ethical given the unique biological communities that thrive around them. Balancing economic gains with the preservation of these unique ecosystems is paramount.

The economic implications extend beyond direct resource extraction. Ocean floor mapping, essential for identifying potential resources, has broader applications. Improved maps of the seafloor are crucial for laying submarine cables, improving navigation, and understanding ocean currents, all with significant economic implications for international trade, communication, and climate modelling. The data gathered during these mapping expeditions can indirectly contribute to other sectors, generating further economic benefits.

However, concerns about the environmental impact of deep-sea mining cannot be ignored. Disturbing the seafloor can cause damage to benthic communities, leading to habitat destruction and biodiversity loss. The release of sediment plumes during mining operations can also have negative effects on surrounding marine ecosystems. A thorough understanding of these impacts and the implementation of stringent environmental regulations are crucial to ensure the long-term sustainability of deep-sea resource exploitation. The economic benefits must be weighed carefully against the potential environmental costs. Failure to do so could jeopardize the long-term health of the ocean and undermine any economic gains. The future of deep-sea resource extraction hinges on striking a balance between economic opportunity and environmental stewardship. International cooperation is essential in establishing clear regulations and guidelines for deep-sea mining. Collaborative research efforts are also needed to develop sustainable and environmentally friendly mining technologies, ensuring the responsible exploitation of these valuable resources without sacrificing the integrity of the deep-sea ecosystem. This careful balance will be key in determining the long-term economic viability of deep-sea mining and its impact on the future of ocean exploration. Without this carefully considered approach, the economic gains may be short-lived and potentially catastrophic in their environmental consequences. It's a complex problem demanding careful attention to all factors involved.

7.5 Environmental Considerations

Our exploration of the deep ocean, while revealing incredible geological wonders, must proceed with mindful consideration for its delicate environment. The extraction of valuable minerals, the potential for pollution from research vessels, and the disruption of unique ecosystems all demand responsible stewardship. We can't afford to repeat the mistakes of land-based resource extraction; the ocean's recovery is slower and more complex.

Sustainable practices must be at the forefront of any deep-sea operation. This includes thorough environmental impact assessments before undertaking any substantial project, meticulous planning to minimize disturbance to habitats and marine life, and robust monitoring throughout the process. Advanced technologies, like remotely operated vehicles (ROVs) equipped with high-resolution sensors, can aid in this process. They can survey the seabed with minimal disturbance, allowing for detailed mapping and sampling without causing significant damage to the environment.

We need stringent regulations and international cooperation to protect the ocean's vulnerable ecosystems. A global framework for the responsible exploration and extraction of deep-sea resources is crucial; one that balances economic development with environmental protection. This framework must include clear guidelines for waste disposal, pollution control, and the prevention of habitat destruction. Independent monitoring and enforcement are also vital. Without strong oversight, the pursuit of resources could irrevocably damage these irreplaceable ecosystems.

The unique biodiversity of the deep ocean demands specific protection measures. Hydrothermal vent ecosystems, for instance, host remarkably adapted organisms found nowhere else. Their sensitive equilibrium is easily disrupted, with potentially devastating consequences for these unique species. Therefore, exploration and exploitation in these areas require exceptionally cautious planning, with extensive research into the potential impacts on the ecosystem before any operation commences. Preservation should always come first.

Ocean acidification, driven by increased atmospheric carbon dioxide, poses a significant threat to deep-sea environments. The absorption of carbon dioxide by seawater lowers its pH, impacting shell-forming organisms and altering the chemical balance of the ocean. This ongoing change affects all marine life and should inform future research strategies, especially emphasizing the study of ecosystem resilience and adaptation to acidification. We must understand these impacts fully to plan mitigation strategies.

The long-term effects of deep-sea mining are still largely unknown. While technological advances allow for more precise extraction techniques, the potential for habitat destruction and pollution remains a significant concern. Furthermore, the dispersal of sediments from mining operations could have far-reaching impacts, affecting ecosystems far beyond the immediate mining site. This requires comprehensive long-term studies to evaluate ecological recovery and the effects on the wider marine environment. Precautionary measures should be considered until a better understanding of these effects is achieved.

Education and public awareness are key to ensuring the long-term sustainability of deep-sea exploration. By increasing public understanding of the fragility of deep-sea ecosystems and the importance of responsible resource management, we can foster a broader commitment to conservation. Supporting educational programs and outreach initiatives designed to engage young people in ocean science and conservation is essential for building a future generation of informed and responsible ocean stewards. Incorporating environmental responsibility from the outset is paramount to success and the future of ocean exploration.

Responsible deep-sea exploration requires a holistic approach, integrating scientific knowledge, technological advancements, and ethical considerations. We must approach our work with humility and respect for the delicate balance of the deep-sea environment. The long-term health and biodiversity of our oceans depend on it. Only by meticulously researching and implementing sustainable practices can we ensure the continued discovery and appreciation of the deep's delicate delights for generations to come. Our obligation extends beyond mere exploration; it's a commitment to preservation for the planet's future.

www.ingramcontent.com/pod-product-compliance
Lightning Source LLC
LaVergne TN
LVHW022014200926
848481LV00007B/2381
* 9 7 9 8 3 4 9 4 1 3 1 5 5 *